YOU
BLACK BOOK OF BUILDING BASICS

COMMERCIAL REAL ESTATE CONSTRUCTION & DEVELOPMENT

JEFF P. MANNING

Book Cover and Interior Design by Daniel Middleton
www.scribefreelance.com

Set in Abadi MT Condensed Light

ISBN: 978-0-615-35289-3
LC Control No.: 2010922114

Published and Printed in
the United States of America

LEGAL WAIVER

and convenience to readers. Such links and references are not intended to, nor do they constitute, an endorsement of the linked materials or the referenced organizations. The content and views on such links and of such organizations are solely their own and do not necessarily reflect those of the author, organization, publisher, or the organizations' officers, directors, or associates.

Illustrations or stories naming fictional characters based upon prior experience of the author are not intended to, nor do they, reveal the actual identity of any individuals, companies, or other entities; and we waive and disclaim any liability associated with sharing these experiences or if the shared experiences sound similar to any given entity or circumstance. Many of these illustrations are based on a true story or are fictional for the purposes of illustration. These illustrations are for edification purposes of the readership only.

In no event will any of the above listed entities, including the author, publisher, or any of its officers, directors, or associates be liable for any damages—whether direct, indirect, special, general, consequential, for alleged lost profits, or otherwise—that might result from any use of or reliance on these materials.

Contents

INTRODUCTION

"A tool to make your development & construction project a success"

THANK YOU FOR READING *Your Little Black Book of Building Basics*. The purpose of this brief handbook is to help demystify and ensure that you understand the building process in everyday language. This is an industry full of acronyms, slang, professional engineering terminology, building codes, and everything in between. Don't get caught in the middle! Knowledge is the difference between success and failure, and with a little knowledge—instead of living a nightmare—you can make your project a fun and rewarding experience.

In this book, we provide an overview of the entire process, highlight the key players, and explain the Building Basics that will arm you with the knowledge to effectively understand the basics of the commercial construction process before you begin. You can utilize this information as a guide as you continue from project conception to your move-in date.

Some of this information is admittedly dry. Yes, discussing engineering and building permits can put you to sleep. ZZZZZZZZZZ . . . So, feel free to use the table of contents and tab right to a topic you need; or, alternatively, read this at night if you have insomnia! We believe this short guide will be fun and informative . . . really.

There is enough knowledge and topics of discussion in each discipline of development, design, engineering, contracting, and project delivery that volumes of exhaustive material are available online, and doctorates can be achieved at today's universities in any of the disciplines, including construction management. This work is by no means meant to be an exhaustive or in depth approach. Phew, that paragraph alone was tiring. . . .

Our goal is to provide you with a simple, easy to follow, guidebook that will allow you to easily and effortlessly understand the construction process and enable you to develop the best team of project delivery organizations.

Our goal is to help you achieve your project on time and within budget, and allow you to set realistic expectations so you can truly enjoy both the creative process and the outcome.

Many contractors say, "Price, Schedule, Quality. Pick Two." I truly believe that the market dynamics have changed, and in today's economy (and within reasonable expectations) customers should have, and deserve, all three—a great price, a schedule that reflects a sense of urgency, and a quality product. My hope is that this book will help you to be knowledgeable enough to demand those project requirements.

Remember, a construction project is a dynamic, living process that involves teams of people, materials, and documents that all work together with momentum and daily action to create a final building that is useful to the immediate owner, the community, and the overall environment for years to come.

Building has been the passion of pharaohs, emperors, presidents, entrepreneurs, and mankind since the beginning of recorded history. Now let's get to work building your kingdom!

As your author, my wish is that this book will provide you with the building basics to enhance your knowledge of the development and construction process, and provide you with a useful resource for years to come. Let's make it happen.

Truly Yours,

—**JEFF P. MANNING**

Your Project

Here we are, ready to embark on a building project. Grab your hardhat, your hammer, your pen, your brain, your wallet, and your patience. Put a sense of humor in your toolbox, take a few deep breaths, and let's get it started!

Tenant Improvement

Often called TI's in the industry, a tenant improvement is a typical interior build out or modification of a space to accommodate the tenant's needs. If you are the landlord, you will want to ensure that the general contractor selected by the tenant is competent enough to ensure that the space is delivered to the tenant correctly and in a timely fashion. What is especially important is that no damage be caused to the existing building or disruption to other paying tenants.

Landlord Rules

As a tenant, your lease agreement should include an exhibit attached or language therein that communicates the landlord's rules for construction operations. It is imperative that a copy of this document be given to the general contractor that will build out your tenant improvement, and your general contractor should include these rules in all of their subcontract agreements so that every company involved in your project will understand the landlord's rules. Typical Landlord rules should spell out on-site working hours, staging areas, trash dumpster locations, security contact information, daily clean up of construction operations, any construction security or damage deposit requirements, permit requirements, any landlord required subcontractors (typically fire alarm and fire sprinkler subcontractors), and, most importantly, insurance requirements for on-site operations. Always check with the landlord to verify whether or not they are carrying a builder's risk policy for the entire building. We will expand on the importance of this policy under Insurance.

CCRs

Whether you have purchased a building within a larger commercial subdivision or you are a tenant, modifications to the building, remodels, and physical changes are usually subject to compliance within the CCRs and may be subject to the review and approval of an association, a board, or a compliance officer. Acquiring a copy of the CCRs is very important in your decisions and is beneficial for your bottom line.

For example, I heard about a story regarding the importance of CCRs coming from a contractor in Southern California. Their company undertook a project in which a business purchased a suite in a commercial project in California. As a requirement for their day to day operations, their client's design called for the installation of an emergency power generation system. This particular client already completed the purchase and closed escrow on the building without reviewing the CCRs to find out if the generator was allowed on property and, if so, what its requirements were. While reading it here seems elementary, this was a large firm with layers of management that provided a service that required everyone in the firm to be highly intelligent, motivated, and in possession of a college degree at the bachelor's or master's level. When the general contractor's staff was designing the project, their project manager informed the customer that they had to meet with their association and attain approval of the location of their generator. The firm didn't really have one point of contact within their company assigned to overseeing the project on their behalf (we will also discuss the importance of this assignment for the owner in another section: Get a POC!).

At any rate, the client left the meeting, and (as was later found out) they never checked with the CCRs. The general contractor had provided them with estimates showing the location of the generator as close to their suite as possible to save on electrical conduit costs and for ease of access. Prior to the construction, in fact, the project manager then had to personally take this simple request to the highest level of their organization to ensure that they checked through their CCRs and acquired a copy. Sure enough, they had to get the location approved, and, upon final approval, the generator had to be located in a different area requiring a lot more work and resulting in a large increase in their project budget.

To avoid this kind of headache, we urge and advise our customers to review their CCRs in detail (hopefully before they sign a lease or buy a building); and that they give these CCRs to their architect and contractor or the design-build contractor. Doing so will save the entire team a lot of headaches and save your clients a lot of money.

KNOWING YOUR LEASE

You've heard the famous saying, "Don't sweat the small stuff;" we advise the opposite when it comes to knowing your lease, lease-purchase agreement, sale agreement, or any legal real estate transaction. There are several resources available to understand leases, and we recommend working with a competent broker that carries professional designations such as CCIM or SIOR. These associations provide designations only to people with years of competency, experience, and education, and they provide testing and, oftentimes, even income requirements for membership to ensure that the broker is of the highest competency. For more information, we suggest you go to www.ccim.com or, if your project is an office or industrial space, visit www.sior.com. Work with a professional and drill down into the details. Understand what you are agreeing to and who the landlord is and what you can expect. I don't ever enter into a lease agreement with a landlord who is not willing to provide me with a review of their financial condition and financial statements. You must understand that they are financially sound and can service you. I also interview other tenants to get an honest opinion of current tenants' relative satisfaction. In retail, you can get some very honest feedback on traffic counts and sales volume. If you can get some honest answers, you will save money.

Also determine what type of service comes with your tenancy. How often are maintenance items addressed at your building and in your common areas? How financially secure is your landlord? What happens in the event of a sale of the landlord's holdings or a bankruptcy? Do you understand utility costs? Are you paying for all your utilities (often called triple nets, of which a common symbol is NNN), some of the utilities (modified gross) or does the landlord pay for all the utilities (full service)? Other things to watch out for are common area maintenance charges (also called CAMs). Please work with a professional real estate broker who is competent in explaining all the terms and conditions of your lease and is working in your best interest. If your broker is really anxious to get the deal done, you may want to re-think utilizing their services. Do not feel rushed. Once you execute the lease, you are locked in with that landlord and their terms and conditions for a long time. Weigh this decision carefully.

TENANT ALLOWANCE

This is the sum of money granted by the landlord/developer to pay for a portion of modifications that a tenant may require. It is typically a per square foot allocation. Tenant allowances range widely from area to area and change with economic conditions. They also vary with venue and market sector. A competent broker should be able to provide a comparison or matrix of typical tenant allowances by product type, market sector, venue, and market timing. In a heated market with little vacancy and high demand, expect the tenant allowance to be low; and in a down market or recession, expect a higher tenant allowance.

Understand at what point you, as a tenant, will receive the tenant allowance dollars. It is also not uncommon for landlord/developers to review your tenant needs and provide the actual modification on your behalf within a certain dollar limit or specification rather than provide a cash allowance. Landlords tend to provide turn-key spaces in lieu of tenant allowances when they feel they have a team in place that they know and trust to provide the design and construction services, and when they feel they will reduce their costs and deliver the project in a more timely fashion than if the tenant is left to do so on their own. Either approach can result in a good outcome for both parties.

CONSTRUCTION TIMELINE

Your lease may also dictate the amount of time you have to complete your construction. It is imperative that this information is shared with your development team. Your design-build contractor should understand your desired move-in date as well as any consequences in the form of liquidated damages or other damages you face for missing an opening date. In addition, the contract between the general contractor and your company should very clearly address the cost to the contractor for late delivery of your project. With that said, it is important that the language is clear and fair, as there are many items that could impact your opening outside the control of the general contractor that you, the owner, must manage.

A competent contractor should provide you with a critical path method (CPM) schedule that is specific to your project. Each and every task needed to complete your project from the beginning to the end should be incorporated into your contract with the general contractor. Delays, slips, or gains to the schedule should be updated as they happen, and a weekly schedule update should be provided for you on a timely basis. You need to inform your landlord or open the lines of communication between your general contractor and landlord so that your general contractor can effectively keep the landlord updated on your opening date. We will discuss CPM schedules in detail in a later chapter.

NEW TENANT IMPROVEMENT

A new tenant improvement, or building out the interior of a new grey shell, is likely the easiest construction project in the commercial market. It's like starting with a clean slate. Important things to acquire are the original shell building drawings. If you can get these digitally, or in AutoCAD (CAD), that is even better, since it will save time on the design for your new space. You should also understand the condition of adjacent spaces (are they occupied?) and understand locations of utilities into your space. With that, your design team will measure and field verify the existing space by measuring interior height, length, and width, and you're off to the races to start your design and construction!

BASIC BUILDING BIT: *Remember to perform within your CCR's, lease agreement, and Landlord construction rules!*

Construction Contracts

Let's briefly touch on the importance of understanding and having a proper construction contract in place. The contractual agreement between your company and general contractor should be well understood by both parties. Too often, I see project managers that do not or have not read or understood their contract, and often the owner does not either. The contract is the basis of the relationship, and it should be very specific. Since real estate and construction law is another exhaustive topic, wherein attorney involvement may be necessary, we will provide you with only the very basics.

Your construction contract should at a minimum define: all the parties and their various roles and responsibilities, the Scope of Work (what is to be done), the cost/price of the contract, timelines and address changes in the work, failure to perform the work by the parties, dispute resolutions, safety, punch list, payment terms, subcontractor(s) roles, access to the site, insurance, hazardous materials, testing, and even termination.

To save time for both parties, an industry standard contract format is highly recommended. The AIA (American Institute of Architects) has produced the industry standard contracts for the construction industry for over a century. The first AIA contracts were published in 1888. These contracts have been tested over time, and there is substantial case law history for these agreements.

While there are a number of different contract formats, most projects will be executed under an AIA A101 agreement, which is a lump sum price agreement. In other words, you and the contractor have an agreed upon price for the project which is the price you are going to pay for the scope of work defined in the contract, outside of any changes.

On small projects, a smaller agreement format currently being used is the AIA A107, which is an abbreviated version of the AIA A101 that is more tailored to projects of small dollar amounts (i.e., less than a couple million dollars).

There are other formats depending upon your project delivery method, such as cost of the work plus a fee, guaranteed maximum price, etc. There are over 80 formats available through the AIA; however, the AIA A101 and A107 are the most used. You will find contact information for the AIA in the resource section at the end of this book.

The AGC (Associated General Contractors of America) also has, in the past several years, established standard contracts similar to the AIA contracts that are also an industry recognized standard.

To keep our discussion on contracts brief (again, zzzzzzzzzzzz, no sleeping allowed!), we reiterate that you should use one of these two organizations' standard contracts, with some modifications made where necessary to the satisfaction of both parties and the project's specific nuances. If you feel it is needed, have an attorney who specializes in construction contracts review the contract prior to execution. The nice thing about using an industry standard contract is that attorney involvement probably really isn't necessary, but we aren't providing legal advice in this book!

BASIC BUILDING BIT: *Don't sign a one- or two-page contract or a simple proposal with a GC! Be very afraid of short contracts and poorly prepared proposals as they leave items undefined that can end up in courts. You want a several page contract that is definitive! An abbreviated AIA A107 should be around 16 pages; others can be 50 pages or more. The more that is defined in the contract, the better protection it will offer both parties; and it will clearly define roles, responsibilities, expectations, and consequences. Ensure that you use an industry standard inclusive agreement that you understand. Your contract should address as many "If-then" and "What-if's" as possible. The AIA agreement has been performing this role for over a century in the US; and a great general contractor will already have these documents ready to go.*

Remodeling an Existing Space

Demolition: Let's Rip and Tear!

An existing space can be a huge benefit because, oftentimes, there are existing items that you can incorporate into your new design that will save you time and money. Important things to understand are the age and condition of equipment, such as HVAC units, kitchen equipment, electrical panels, and restroom fixtures. Hopefully, you have done a complete review and had the building inspected prior to your lease agreement execution or closing. In a later chapter, we discuss Existing Conditions and the importance of "discovering them before they discover you." If your new space requires a lot of remodeling and demolition, we urge you to perform the demolition as soon as physically possible.

Current Code vs. Existing Conditions

Reading building codes is like reading the tax code; it will cause most people to go into a daze that lasts for days . . . daze . . . or what was that? Just mentioning code makes sentence completion difficult!

Building codes are constantly being changed, and, with each passing year or update to the building code, they become more stringent. The building code requirements vary from area to area, depending on variants such as environmental conditions, seismic zones (i.e., requirements in heavy earthquake prone areas such as California are much different and more stringent than areas that do not experience earthquakes), etc. Sometimes, code interpretation is subject to heated debates. Building codes are also amended at the local and state levels based on past tragedies.

For instance, in Nevada, at the time of this writing, all low voltage wiring is required to be installed in a conduit (a rugged, protective tube through which wiring is pulled). Many other states do not require low voltage wiring to be in conduits because they are typically coated and do not carry a large current load. However, several years ago the MGM hotel caught fire and several floors were burned. It was a huge blaze and a real mess. Not long after, it was mandated that all low voltage wiring in the area be installed in conduits, as it was deemed a fire hazard since authorities traced the MGM fire to a low voltage wiring issue. Now when owners and developers come to Nevada to build, they are usually amazed at the high cost of low voltage systems. It is because the systems must all be installed in conduits and inspected. If your existing conditions contain old wiring not in conduit, the building department will require it be corrected to meet the new code.

Understand if the age your space warrants an Asbestos Abatement or Hazardous Materials study. This should be recognized prior to your lease and during your due diligence. These items have serious health and cost implications, and authorities with jurisdiction, such as the air quality management agency or building department, will require an abatement survey and removal of their existence prior to issuance of a building permit for your remodel. Discuss asbestos with your real estate broker prior to leasing a space, and certainly discuss it with your general contractor.

Existing conditions pose huge budgetary and schedule issues to a new tenant or building owner. It's very important that due diligence is conducted to determine what issues need to be addressed.

BASIC BUILDING BIT: Do all you can to understand if there are hidden costs at your site by uncovering issues upfront that could cause unforeseen delays later—As-builts of an existing building, early demolition, potholing the earth for underground, and discussing options with consultants or your experienced general contractor!

NEW BUILDING

In a new building situation, the accuracy of the civil and utility drawings is paramount. Typically, the largest risk is posed during the site work, and can also arise from utilities and off site items that need to be considered, correctly engineered and estimated, and installed both correctly and timely. The fact that we cannot see underground poses the biggest financial threat to the project. Careful pre-planning with all utility companies for locates of existing utilities and correct civil drawings with proper drainage and flow considerations are just a few concerns. In addition, the contracting team should be diligent in dust mitigation, knowing the process, timing, and requirements of air quality management authorities, and SWPP (Storm Water Pollution Prevention) plans to mitigate sediment from your site before it pollutes the environment.

Another concern you must consider is site security. As the owner, ensure that a Builders Risk policy with specific theft and vandalism riders are purchased with a low deductible that makes sense. Also, depending on the area, utilize a temporary site fence, install motion sensors, and even consider employing on-site security during the hours that the construction site is not in operation. Having personally experienced coming to a job site to start activities on a Monday morning only to find broken windows, stolen materials, and missing computers from the on-site trailer, I can assure you that proper insurance for theft and consideration of security for a new building project is an absolute must. Later chapters of the book will discuss project management and considerations for your project regardless of it being a remodel, a tenant improvement, or a new ground up structure.

HONESTY, COSTS & MARKET VOLATILITY

Current economic conditions will determine the availability of resources for your project, and markets will dictate the costs. When working closely with your general contractor and architect, you will also need to determine a realistic budget and schedule that will meet your needs. Do not penalize your potential general contractor for being honest.

For example, I once met with a new customer who was the CEO of a publicly traded firm that wanted to build a medical office. Let's just call him "Gary." If your name is Gary, please don't be offended. Gary was an intelligent individual with his medical degree and in the top of his very specialized surgical field. He had high expectations for his design. Gary wanted several custom elements including a water feature.

I informed him I was concerned about his budget. He told me that he still wanted his laundry list of custom features. After reviewing his space plan—which was over 10,000 square feet— it required extensive demolition of existing interiors, a new high-end foyer and reception area, extensive "all glass and metal elements," several laboratories, a day surgery area, and some X-Ray rooms. At the time of the estimate, I honestly told him to expect the project to be well over a million dollars. I also told Gary we could work with him to design his space on a budget within certain limits if he was willing to make some concessions on the extent of his design. He told me that he had a tenant improvement allowance of $350,000. He even called the person who referred us and complained that "the contractor's prices are too high."

Two days later, Gary emailed me saying, "I found a friend of mine that happens to have a contractor's license and has assured me he can build it all for $375,000." I wished them the best of luck. How a person who happens to have a license can promise a budget with no design was beyond my ability to rationalize. My expertise at the time included seven years of medical and laboratory projects, both at the local level and across state lines, and twenty years of construction experience, and our company was staffed with over a century of combined experience.

Who would you listen to? Sometimes the advice you get may not be music to your ears, but you should always put stock in the advice of people based on their résumé, community references, professional designations, and even through a referral; and you know that their answer is the truth. Unfortunately for Gary, someone else told him what he wanted to hear, and he will pay the ultimate price of a budgetary bust, delays, change orders, and the headaches associated with hiring unqualified contractors who lie. Avoid doing this.

I see intelligent, sophisticated people every week who are just like Gary—they are leaders in their respective fields, yet they constantly hurt themselves on projects because they really do not understand building basics. They underestimate the huge financial risk they are undertaking and do not understand the importance of proper contractor selection and engineering. They do not understand costs, building codes, or timelines, and yet they will hire anyone who just tells them what they hope to hear or, worse yet, "get bids and hire the lowest."

BASIC BUILDING BIT: *Please, make me a promise: You will hire someone that will tell you the truth. I always give "the bad news" first and then we manage it to the best outcome. Ensure that your contractor does the same thing. I am glad you are reading this book so that you won't end up like "Gary"!*

The Project Team and Your POC

THIS IS THE MOST IMPORTANT decision that will be made concerning your project. An overall project team can vary by stage and scale of a development and could include the equity partners, financier, real estate brokers, project managers, contractor, architect, several engineers in various scopes, consultants, subcontractors, suppliers, public relations/marketing personnel, third party voucher control services, utility companies, and community members, etc.

Let's assume you are a private owner, capital improvement manager, tenant or landlord on a typical project that you are either building from the ground up or building out from an interior (TI). The team consists of you, your in-house project decision maker or project manager (which may also be you), your contractor, the architect, sub-consultants, subcontractors and suppliers, and, if you are a financing the project, your financier. The largest mistake most companies make when starting a commercial construction project is failing to define roles and failing to appoint and define one single point of contact (POC) in-house that is in charge of the day to day decisions of the project on behalf of the owner. Always get your POC established!

INTERNAL PROJECT DECISION MAKER – A KEY TO SUCCESS OR FAILURE.

You must realize that as the owner, even with the most competent team of contractors, architects, engineers, and consultants, timely and important decisions regarding the scope of work (what will be built) and the budget (how much it will cost?) are ultimately your responsibility. A great design-build contractor will supply you with options, honest cost data, and interior design choices, but decisions can only be made and approved by the owner. Most owners often have multiple persons within their own company that have a stake or need to provide input on a project.

For example, say a group of doctors purchase a building and want to build out a medical practice. Each doctor is very busy managing their existing business and the day to day demands of running a medical clinic. The current practice must continue to operate; yet now the doctors have taken on a new construction project, likely an area in which they are not experts. It is very important that the group meet internally and assign one person in their group as their final decision maker and point of contact for all communication to the contractor, architect, sub-consultants, etc. This person, whether in-house or outsourced, should, at a minimum, understand the timeline for the project, as well as the needs of each stakeholder, the contract, the plans/specifications, and be the driving force on the "owner's team."

We recommend that a general contractor be selected early on in the project to work closely with the owner's main point of contact, and, preferably, design-build the entire project. Decisions need to be made timely and accurately. If you hire a great design-build contractor, they will likely have a person on their team who, as a value-added service, can really be a huge resource to work with various people in your organization; yet, to avoid confusion and delays, the owner should have one point of contact that is the final decision maker and has achieved internal team congruence on your project.

There will be items outside the authority and scope of the contractor or architect that a point person with authority from the owner's team—who will act on their behalf—must initiate. For example, submitting applications for utility company accounts to be established, signing and approving invoices, paying requests, making modifications to the project scope, communicating with the finance company, choosing final colors and finishes that are presented by the designers, conducting final review of the space plans and layout of the project, hiring a third party QAA (Quality Assurance Agency) firm if required by the local building department. None of these decisions can be made by the contractor or the architect—no matter how turn-key or competent their services may be. A point person on the owner's team must be established, and they must be available and have the authority to make timely decisions in order to meet the project timelines.

As an example of what to not have occur on your project, our company provided a complete turn-key design-build project to a business services client that was privately owned by several partners. We initially met with one partner. A contract was established, and we completed the full design and construction documents. We were told via email and a phone call that the plans had been approved by the partners. Upon receipt of the building permit we began construction. The partner who was our original point of contact was very pleased.

Three months into construction, after all the walls were framed and the drywall was installed and painted, the partner came to us and informed us that he had left one of his four partners out of the decision making process because he was on an extended trip and had "trusted him to handle the project." Apparently, the fourth partner came into town and our point of contact had left out many of his wishes in the project. This caused a re-design, an 18-week extension to the schedule, and a change order to their budget for added work items and extended general condition costs. Our company even absorbed some of the costs because of this mistake in order to "help their budget."

Our goal is to ensure that projects run very smoothly on behalf of our clients. However, the owner failed to achieve internal congruence and dictate the needs of the project accurately. It cost them a lot of time and money, as well as a delayed opening; and it made the entire project experience feel negative. It is very important that congruence is achieved internally within your group or organization, that the one point of contact in charge of your project discloses all the stakeholders to the design team and contractor, and that each person with a stake in the project reviews both the project and the plans and even signs the plans. Do not be like the customer who failed to disclose his stakeholders to the construction and design team and caused themselves additional costs and delays. I am glad you are following our advice and are achieving internal congruence.

CONTRACTOR

Often a bad word associated with what people consider a stressful event, the term "contractor" is tossed around very loosely. Contractors come from all walks and there are many levels of contractors, from Joe in his pickup truck, who may do handyman services and/or deck repair, to the publicly traded, sophisticated firms, and everything in between. This is the single most important decision you will make on your project (along with choosing an architect, if you are hiring them separately).

Think of choosing your contractor as choosing a short term marriage partner. If the project is going to take a month or two years in duration, you will interact with this contractor on a weekly basis for the duration of the project. When it's ugly, you are still connected by the executed contract, and getting a divorce from your contractor mid-project can cost a fortune. I have often been called into projects that have been turned upside down by incompetent general contractors who did not have the expertise or resources for the type of project that they pursued; and as a result, the owner ends up in a legal battle while running out of time and money and has a half-built mess that someone has to clean up. Creating buildings in the physical environment is not easy to correct when done wrong. We can't roll up to the project site and hit "delete," create a blank page, and start over. It's so important to get competent people onboard the first time. Doing your due diligence and selecting a contractor correctly based on a number of important factors is an absolute must.

Basic Building Bit: *You're about to get married! Get to know your contractor. Reference, Résumé, Google them! Do your homework!*

WHO IS THE CONTRACTOR?

The general contractor (GC) will be the single most important driving force on the project. They will be responsible for the final outcome impacting the cost, scheduled delivery date, quality of the project, *quality of the experience you live through*, and warranty after the project.

WHO IS ACTUALLY BUILDING THE PROJECT? WHAT TO LOOK FOR?

Licensed, bonded, insured—a no brainer! This is the obvious, and many companies have achieved licensing, bonding and carry insurance. We won't belabor this point except to note that you should remember to double check and make sure that they have these legally required items in place.

Expertise. It's important that you hire a contractor that is truly an expert in the type of building project you are planning. While this seems obvious, time and again, we encounter clients that hire their residential homebuilder friend to construct their medical office or their distribution warehouse, at a huge cost to themselves. I am aware of a recent eighty million dollar project for which a very intelligent CEO hired a residential "buddy" to oversee the project management. Can you imagine the nightmare and cost that this CEO will experience? Again, hire an expert. Determine their expertise by résumé, and most importantly, references. A real construction expert in today's market will have built such a name and reputation that you should be able to "Google them"—type their individual name and city into Google and see if they have made positive headlines for several years. If they haven't, perhaps they aren't the expert they claim to be. In fact, ask them, "What would I find out if I were to Google you?" LOL.

STAFFING – WHO?

Ensure that your contractor has assigned adequate resources that will meet your project needs. Are you getting principal level involvement? Will an owner or senior manager of the company be involved in your job? How many projects does the firm have going on and where does your project, by size (i.e., dollar cost), fit into their mix? Will you get their "A" team resources or not? Ask them, "Who will actually build my project?" and ask for their résumé. Ensure that you get a top down commitment on exactly *who* your potential GC firm will place in charge of the project.

When I consult owners on contractor selection, I encourage them to choose GCs that have the resources to handle the project. Also, your project should be large enough in their eyes that, even given their overall workload, it is important to the people at the top. Otherwise, you risk getting lost in the shuffle and being just one of many projects at a firm, and you're left hoping that you get a good project team that stays employed with that firm long enough to see your project through from beginning to end. I would rather be one of a smaller firm's larger clients, making my project a big fish in a smaller pond, than hire a huge general contractor and have their lowest talent on my job. Make sure you do due diligence on the staff and get a commitment from senior management in writing. For instance, in construction contracts that I am involved in, we state, "A principal or Senior Level manager from our firm will be involved, on call and available to manage the day to day details of your project."

RESOURCES

A general contractor's resources are its people, relationships with subcontractors and suppliers, relationships with building officials and utility companies, and its financial capacity relative to the size of your project. It's pretty obvious that choosing a contractor that shows up to your initial meeting coated in drywall dust with a tape measure on his waist and his pickup in the front of your office may not be the best choice. They will always be the cheapest, but they will also likely lack the resources to really deliver. Choose a contractor that can demonstrate that they have the horsepower to complete your project, and that they run a professional organization. The lowest bid and the cheapest price will not buy you a professional; it will buy you a nightmare!

BASIC BUILDING BIT: You wouldn't hire a taxicab driver to fly you and your family to Europe on a 747! Crash! Choose a contractor that has the project expertise, and make sure that the upper management and owners demonstrate their ability—via their résumé of previous projects—in building a similar or more complex project than yours!

VERIFICATION

How do you choose a contractor with resources and project expertise? Interview and verify.

Within most states there is a State Contractors Board or licensing entity that oversees contractors. You can contact the board with the license number of the firm you are considering and you can find out if, and what types of, complaints have ever been filed with the board.

Call their references. Most contractors should be willing to provide you with a list of references. It doesn't hurt to call them and ask about their experience:

Was the project handled professionally?

Did the contractor give the past owner a lot of change orders that the owner did not request?

Did the contractor deliver the project on schedule?

Was the job a "safe" project, meaning no accidents?

How was the level of quality?

Have there been any warranty items since completion? If so, has the GC been fair, fixing any warranty items on a timely basis?

Who was the past customer's main point of contact with the GC?

Were they satisfied and would they use this GC again?

Another way of verifying a contractor's reputation is by looking into their level of industry trade association and community involvement:

Are they members of the Associated General Contractors?

Are they members of other groups such as NAIOP (commercial real estate development), SIOR, CCIM, or others?

Does this company ever contribute to the community charities?

SUBCONTRACTORS

A subcontractor is likely someone that you will never deal with directly—as long as you have hired a good general contractor. OK, the definition of a subcontractor is a secondary contractor that has a contract with the general contractor to provide or perform specific trades on a project that the general contractor does not self-perform. These are usually referred to as "specialty contractors"—i.e., they specialize in electrical work or plumbing only.

The most important thing an owner needs to do is to get a commitment from the general contractor who has properly pre-qualified and hired reputable subcontractors, has received competitive bids from the subcontracting community, verified that prices charged are competitive, and that, prior to paying, the subcontractors' work is installed and proper lien releases have been collected. It's important that quality subcontractors who can deliver their portion of the project on schedule are chosen. They must also install quality workmanship, have the resources to handle the project, and hopefully be in business for future follow up warranty work. A great general contractor is as careful about which subcontractors they hire as an intelligent owner is in choosing their general contractor. Subcontractor selection greatly impacts the entire project team.

"Cheap Bob" is an owner who wants to pinch pennies. I agree that we all want to attain a project delivered for the best possible price. Cheap Bob did one thing right; he was smart enough to select a general contractor to bid the project out to several subcontractors on his behalf and negotiate it rather than going out and just getting "three bids" from competing general contractors. However, Cheap Bob has very unrealistic expectations and just looks at price. He expected his GC to hire any subcontractor that he could find. Cheap Bob gave his accommodating

GC a few names of some electricians that had "great prices." (Unfortunately, there are a lot of quality problems in the marketplace; and there always will be due to the segmented nature of the construction industry.) In order to accommodate the owner, the GC agrees to hire a few untested subcontractors that were not pre-qualified prior to commencement of the project. The GC still has the risk to deliver the building to Cheap Bob. Halfway through the project, Cheap Bob's favorite electrician cannot man the job because he doesn't have enough employees. To make matters worse, he is failing to pay his suppliers on time. As a result, the project is behind schedule, and the lighting supplier will not ship any light fixtures to the project. The GC is now stuck trying to negotiate with the lighting supplier and must hire another electrician, with delays and at additional cost, to come in and finish the project. Cheap Bob is unhappy, the GC who was trying to accommodate Bob is unhappy, and, to top it all off, the lighting supplier is going to lien the property. The job is now behind schedule and over budget. Don't be like Cheap Bob!

Our answer to Cheap Bob is hire a reputable general contractor that has a database of pre-qualified, reputable subcontractors, and allow the GC to hire subcontractors that they know will deliver the project on time, within budget, and without these hassles. Do not chase pennies and end up losing dollars! If Cheap

Bob can do a better job selecting subcontractors and dividing up work scopes, then he doesn't need a general contractor. He needs to get his own license, and he can try to build it himself.

BASIC BUILDING BIT: Never direct subcontractors and always go through your general if you want anything communicated to a subcontractor. Don't force your friends who happen to "own a drywall company" on your general contractor. Keep business decisions separate from your personal ones and your project will go much smoother. You have hired a professional general contractor to deliver your project to you, and shift your risk and management headaches so you save time and money in the long run.

ARCHITECT

Your architect's role is really to design the building to the specifications of the client while ensuring that proper building codes are followed and the final product results in a functional, aesthetically pleasing design that adds value to the long term built environment. The architect is often charged with the administration of the project on behalf of the owner. Often, architects work diligently to get plans approved by the local authorities within their jurisdiction, including planning, zoning, the sanitation district, and the building department. You may hire an architect directly or you may hire a design-build general contractor. A very popular approach in today's environment is to choose a general contractor that either has an in-house architect to provide architectural services or one who consistently outsources their design needs to a state licensed professional.

At a minimum, the architect should be a member of the American Institute of Architects (AIA), be licensed in your state, and again have expertise in the size and type of project you are undertaking. Do not hire a residential architect to design your commercial project. You will have a nightmare on your hands due to vast differences in building codes and the complex process of acquiring approvals from authorities having jurisdiction. Even if your best friend is a residential architect, you should hire a commercial professional for your commercial project.

INTERIOR DESIGN

The interior designer (ID) may or may not work for the architect, depending on size. I call the ID people "the artisans." These are the creative people that will assist in space planning and in selecting finishes and colors for your space. On large scale projects, this task is done by a sub-consultant team of interior designers that work for the architect. On small projects, your architect or design-build contractor can handle interior design. Expect to have a number of interior design meetings to review, understand, and approve your space plans and your finish materials (i.e., type of flooring, colors of accent walls, etc.).

MPE, STRUCTURAL

In addition to the architect and/or interior designer, your team will likely have some engineers onboard. Typically, the mechanical system, the electrical system, plumbing, and the structure will all be designed and engineered to meet proper building codes by a specialized, licensed engineer in each discipline. These firms will either report directly to and work for the architect, much like a subcontractor who works for the general contractor, or they will work directly for your design-build contractor. The architect or general contractor will select the best team of sub-consultants and match their size and ability to you and your project.

CIVIL ENGINEER

On a ground-up project, you will have a civil engineer on your team. They will typically work for the architect or your design-build general contractor. There is a lot of risk in this discipline that can add up quickly. It is very important that a thorough understanding of the site is accomplished. It is the responsibility of your civil engineer, who works closely with the architect and/or design-build contractor, to ensure that this is accomplished.

PLAN EXPEDITORS

Depending on how busy the local building department is, acquiring your permit once submitted can be difficult. Simple projects can take 9 months to a year to achieve all approvals, which can be reduced in slow times. However, when times get slow, a lot of building authorities cut staff and the timelines for acquiring permits are just as long. It can seemingly take forever to acquire these approvals. On large scale developments, or even on "simple TI's," achieving the permit can adversely impact the ability to open and start generating revenue. Time is money!

So, one solution is to find out if your design-builder or your architect has a permit expeditor. I often use a permit expeditor in one county who is a former senior building official. It helps that he can review plans prior to submitting them, and he can flag any issues that he sees with them. Additionally, he knows everyone at the building department very well. This helps us get our plans approved efficiently, get timely and honest answers, and has paved the way for a great working relationship between myself and the building officials.

"OTHER VENDORS"—*low voltage, fire life safety, signage, security, point of sale, maintenance*

Besides your design-builder or your architect, subcontractors suited for a particular construction project, and building officials, you will likely need to create a list of "other vendors" that you may need. For instance, audio visual, kitchen equipment, and security companies are all firms you may or may not need to hire depending on your project's requirements. I always tell clients to have their general contractor hire as many vendors as possible so that the project schedule is recognized and met by every vendor, and so that costs and coordination are handled with by a single source.

BASIC BUILDING BIT: GIVE YOUR GENERAL CONTRACTOR AS MUCH OF THE PROJECT RESPONSIBILITIES AS POSSIBLE. THIS LIMITS THE AMOUNT OF WORK THAT YOU, AS THE OWNER POINT OF CONTACT WILL HAVE TO ACTUALLY DO, AND YOU WILL HAVE SINGLE SOURCE RESPONSIBILITY. FREE YOUR TIME TO DO WHAT YOU DO BEST AND LET A BUILDING EXPERT DELIVER THE PROJECT!

Utility Companies

Power!

You need to turn on the lights when you open! Owners often forget to apply to the local power company; and they often forget that if it's a new project, the power company may need to include a design to actually construct a power connection. Chances are there is already a power utility vault in the street or on or near the site; however, conduits need to be buried in the ground and a point of connection and meter set at your building. The owner must ensure that this item is addressed.

I was once brought in to rescue an owner who had forgotten about his power requirements. Their GC had excluded all power company requirements from their bid. This was about a $9M commercial project. The owner ended up being stuck with a change order from their GC for over $200,000.

I always ensure that my firm goes out of its way to point out these "dry utility" issues to potential clients. Power requirements should not be a surprise, and you must include the cost of power design, application, deposit for services if required, and the cost of constructing items the power company may require in order to power your project. Don't get *shocked* on your budget! Include the total cost of power in your pro-forma.

OTHER UTILITIES/EXPENSES

Other common utilities include natural gas, your cable/phone service provider, water district (a water meter in some vicinities is more than $30,000!), and even satellite providers (this may require roof penetrations). Put together a list of all the dry and wet utilities your project requires and ensure that you get onboard with them so it's included the first time.

BASIC BUILDING BIT: Before starting your design, create a list of all utility firms that will service your facility. Ensure that you budget both time and financial resources for utilities and start the application process immediately.

Financial Institutions

Your Loan

How you pay for your project is obviously a concern. From the construction standpoint, your bank will likely require an estimate of cost by a reputable construction firm. Additionally, keep in mind that your construction cost, often referred to as a "hard cost," is only one or a few line items in the total pro forma cost of your project. You should purchase or obtain a pro forma that includes line items that encompass the entire cost of your project, including insurance, marketing, fees by government agencies, fees by utility firms, studies, etc. If the project is large enough in scale, I urge customers to have a third party review their pro forma and ensure that they have captured the true total cost of their project prior to submitting for a bank loan. A well done pro forma will go through underwriting with minimal objections and show your banker that you are an experienced, intelligent businessperson worthy of credit and bank risk.

Bonding

Your financial institution may or may not require a payment and performance bond on your general contractor. You need to understand this requirement because if it is the case, you will need to hire a contractor that is bondable and bondable at a competitive rate. A competitive bond rate is no more than 2% on small jobs, and on large scale projects it usually hovers around 1% (at the time of this writing).

I tend to think that bonding GCs is ridiculous. It adds unnecessary costs to the project, and they are rarely collected upon. Contractor selection is more important than bonding. Even during the recent "great recession," most contractors that failed only failed due to owners and banks not paying them. There were few issues with payment and performance of the general contractors or subcontractors. If the right general contractor that meets the criteria in this book is selected, a bond is additional cost that isn't necessary for most projects. However, check with your attorney and banker regarding the necessity to bond your particular project. Acquiring performance and certain bonds for key subcontractors that do specialty trades involving the building envelope, such as the roofer, glazer, or EFIS subcontractor, is recommended on larger scale projects, mainly to ensure that this work item is protected. Often, as a substitution for a Payment and Performance bond, an owner may want to employ a third party voucher control firm. For more information on this topic, turn to the section called, *Progress Payments*, where we get into more details.

BASIC BUILDING BIT: Interest and Your Schedule – one last word on financial institutions, it's very important to understand the cost of interest carry on your construction loan, and you must ensure that your budget will cover the debt service on the loan. You need a realistic schedule so you can calculate your interest carry. You also need to hire a GC that you know will meet the schedule!

YOUR SITE'S CONDITION

ACQUIRING A SOILS REPORT and understanding it is important. Often, soil boring samples are needed to understand what is below the surface of the site. Soil reports with boring logs are utilized by the civil engineer and the structural engineer, and this impacts the size of the footings, thickness of slabs, and thickness of asphalt in parking areas, etc. During the due diligence process of even purchasing a lot, soils reports are necessary in that they are required by building officials to release permits.

ENOUGH FIRE PROTECTION PRESSURE

During the due diligence period, make sure that there is adequate flow in the water lines to pressurize your building property. If you are very savvy, you can employ a firm to provide a flow test at your site. This will ensure that you do not end up having to purchase any expensive pumps if you are building a multi-story or large scale building. Typically, most sites have adequate flow, but this is something that is often overlooked and can have huge financial implications on a development.

UNDERGROUND UTILITIES: ARE THEY IN THE RIGHT SPOT?

If you can acquire as-builts of existing utilities on a ground up project, you will be ahead of the game. You can also do some exploration by pot-holing (using a small backhoe and/or hand digging to locate exactly where utilities are on the site.)

To do your homework, try to acquire an as-built. An as-built is just what it sounds like. It's a plan that shows what was done in the past, and what should be there - "As it was built." You should always also utilize a "call before you dig" service (this is often required by local ordinances). By making this call, utilities will be marked out with colors painted on the ground that correspond to each utility. For instance, gas will be painted yellow, water blue, sewers green, and power is usually marked with a red paint stripe.

REMODELS, DEMO FIRST, THEN DESIGN

If you are a tenant doing a tenant improvement or any owner performing an interior remodel, I encourage you to acquire an early demolition permit and remove as much of the existing items that are to be demolished as soon as possible during your design process. This isn't always an option due to user constraints or operations or landlord requirements. If and when it is possible, this minimizes surprises and reduces your financial risk exposure. What? We don't have X-Ray vision yet? I wish that I could provide a service to owners that enables them to see through ceilings, behind walls, and under slabs, but we haven't arrived at this technology just yet. In order to understand hidden and unforeseen existing site conditions on a remodel, the sooner the demolition is performed, the better. This will allow your design-builder or architect/engineering team to truly survey the existing space and provide you with an accurate estimate.

For example, in an old shopping center, I requested that we complete the demolition of a design-build space prior to completing design. Good thing. It turned out that on both sides of the tenant space the walls did not meet code. Had this been encountered later in the process, it would have been a total financial surprise during the course of construction and would have added two weeks to our schedule. We were able to uncover this situation first, so we could plan for it. Planning is everything! Discover your site before it discovers you and save time and money!

ASBESTOS

You must assume that any building built prior to 1980 contains asbestos. What is asbestos anyway, and why is it a "big deal?" Asbestos is a naturally occurring mineral fiber that was mixed into over 3,600 construction materials. Asbestos was used because of its amazing properties. It is noncombustible, noncorrosive, nonconductive, and has huge tensile strength. Amazing! There is "white asbestos" and "blue or brown asbestos"—certain types actually have difficult names, such as chrysotile, mysorite. . . Not that the names matter to us! What matters is that it is found everywhere in older buildings, because it was used to fireproof, soundproof, insulate, and even to decorate spray-applied ceilings, etc. Stay clear of it! Unfortunately, even limited exposure can result in a disease called Mesothelioma, and it can cause Asbestosis, which is a serious disease of the lungs. If you smoke, exposure to asbestos can increase your risk of lung cancer by 90 times more than a non-smoker!

What to do? If you're planning on doing any demolition, remodeling, or construction work in an older builder, you must contact an asbestos survey firm and have the building "tested" to see if anything in the building contains asbestos. After that study, you then hire an approved Abatement and Hazardous Materials removal firm to legally abate the building prior to any new construction or before any further demolition can be performed. The EPA, California's DOSH, OSHA, and air quality management authorities will all impose heavy civil and even criminal penalties if you do not handle this issue within the law. Also, if you purchase an older building, ensure that you understand the risk of asbestos during your due diligence period and prior to closing! Asbestos kills people; don't mess with it!

Project Delivery Methods

Design-Build, Single Point of Contact

One of my favorite topics that I couldn't get to soon enough in this little book is design-build. Much like LEED being a part of the future of how buildings will be designed and built, design-build is quickly becoming the preferred method of project delivery. What is design-build? In a design-build scenario, the design-builder is a single point of contact (single entity) that is both the general contractor and design professional. The great advantage to design-build is total team congruence, single point of contact for the owner, mitigation of financial risk for an owner, and it can reduce your schedule by allowing the design-builder to combine items in the design phase and construction phase of a project. For instance, on a remodel, demolition may be performed on a demolition only permit, and the space thoroughly surveyed and incorporated into the design. A design-builder can also provide you, the owner, with estimated probable costs of construction during several phases of the design, and help you select items for the project that fit within your budget. If you establish a realistic target budget with your design-builder, you can rely on them to stay on target.

"Hard Bid"

An old method of contracting is design-bid-build or to "hard bid" the project. I think the appropriate word is "hard;" while this is the approach that most people think of—hire an architect, draw it, and go get bids—it is really the hardest way to project success and it makes an enjoyable experience and positive outcome unlikely. When plans are drawn by an architect, and put out for bids among competing general contractors, the contractors, aware that they are competing, are compelled to "only bid what is on the plans"—exactly that, nothing more, nothing less. If you interview an honest architect and ask them if everything needed for the project is 100% on their plans, their honest answer will be certainly "no." Often, design-intent is communicated on construction documents but not with 100% accuracy. This leaves construction projects subject to interpretation and "hard bids" and their owners wide open for change orders or financial impacts.

Here is a short example. I am aware of a contractor that once built two retail stores that were very close to each other in an existing shopping center. Bill was recommended to the general contractor by the landlord of the center. He had not started his design. He had just leased his space. Bill, based on the contractor's reputation, résumé, and expertise, contracted with them to be his single point of contact design-builder. He then jumped on a plane to another country for about six months. During that six months, the GC communicated with him via phone calls, faxes, and emails in order to give him all the details of his project. The GC was able to survey the existing space adequately. They were even able to incorporate items in the existing space into his design criteria. At two intervals during the design, price checks were provided on the design to ensure that we were tracking well on Bill's budget. When Bill returned to the US, his project was well under construction. He was delivered a turn-key of his project on time, without delays, and within his budget. He opened his store successfully, and he was so pleased with the process that he actually became a friend of the contractor's project manager and wrote letters of recommendation.

Thomas leased the space right near Bill's space. He hired an architect who drew up a new plan. Thomas then put the project out to bid among four competing general contractors. The same GC that built Bill's store had some economies of scale due to already being on site. So, they priced exactly what was on the plans and performed a survey of the space, but it was an existing space with everything hidden. The GC could not do any demolition prior to the design being done. Subsequently, after the GC started construction, they uncovered a huge water pipe going through his space; and uncovered the fact that his plans did not fit the space adequately because they also found additional square footage and even a plumbing fixture that was hidden behind a wall. Construction was halted, and the entire team had to wait for a redesign to be done by the architect. Additionally, getting the water lines relocated so the plan would work in the space took time. Thomas' schedule was delayed by over eight weeks. Although his space was smaller and he started his design sooner than Bill did, his store opened weeks after Bill's store. Thomas was several weeks behind his original schedule, and he had to pay the architect to redesign the space, and pay for the change order cost of the plumbing issues. This example proves to me that design-bid-then-build is the poorest choice of project delivery. Thomas agreed that the next time he builds a store, he is going to negotiate with a design-builder or, at the very least, get the contractor onboard early to work with the design team to mitigate issues that lead to serious financial impacts.

BASIC BUILDING BIT: Negotiate your contract with an honest, reputable design-build general contractor first. Then, allow that single entity to be responsible to deliver your success. Have your GC get competitive subcontractor bids in the market and build the project as your partner. Do not draw plans, get competing GCs to bid off them and expect a headache free project. The old method of delivery—"get three bids and build it" has failed, and that is why the entire industry, including government agencies, are all moving quickly to design-build contracting services. Choose to be a success!

THE NEGOTIATED BID PROCESS

There is a way to have a competitive performance and still negotiate your project. You do this by interviewing a couple of general contractors and have them compete based upon the general conditions and fee they would charge to do the project, and by submitting those two items along with a résumé and project presentation. In this way, you can decide what company will perform the work the best on your behalf and approximately what their fee would be. A fee is typically the final profit that a contractor makes after the subtotal of all direct work, general conditions, and insurances. General conditions are the indirect cost of providing a construction project—i.e., the cost of the on-site superintendent, project manager, contract administration, safety, trailers, and dumpsters, etc. If you can understand the general conditions cost and their final fee percentage mark up, you can set a baseline.

You must be careful, I have heard of contractors stating they will do the project for a 2% fee, only to bury and hide money throughout their direct cost estimates. Typically, the fee depends upon the size of the project and will vary anywhere from 5% to 25%, depending on the size of the job; and general conditions will also be anywhere from 5% to 15% of the direct cost of the work. The average general contractor in the USA returns 7–12% to their bottom line annually, after expenses, for everything involved in the building process. Be happy to pay this small fee and know that you have an expert.

THE IMPORTANCE OF EXCELLENT DRAWINGS AND SPECIFICATIONS

DDs THROUGH CDs (DESIGN DEVELOPMENT AND CONSTRUCTION DOCUMENTS)

It's very important that your design team produce high quality drawings containing as much detailed information as possible. This is the most important tool in the building process, as it must completely communicate where and what is to be constructed. Detailed specifications tell us what must be installed, including product type and manufacturer. General notes should not replace details on a set of plans. Often, designers like to put a blanket statement on plans and not single out specific items. This only leads to a lot of questions, formally referred to as Requests for Information (discussed later), and delays to the project schedule. The more detailed the information is on the plans, the easier the construction process will go. Check your plans to ensure that the information is applicable solely to your project; a lot of architects get sloppy and cut and paste standard details onto your plans and include items that do not apply to your job. This causes a lot of confusion. The plans also need to be community specific, including local building code amendments. They must address the concerns of all authorities that have jurisdiction over your project, including the health department and/or fire department.

Schedules

I WAS SHOCKED TO WALK A 150,000 SF tilt up warehouse project a few months ago with an owner who was in complete frustration. He had made a really poor choice in hiring a friend's company that had typically only built residential projects. They never provided this developer with a CPM schedule! What is a CPM schedule? A critical path method schedule is a tool used in many industries that establishes all the tasks that need to be accomplished on a project from beginning to end, and it identifies the most critical items and links them in a logical order. CPM scheduling can be very complex, but on smaller jobs it is very simple. Every project should have a CPM schedule that shows the start date, long lead items, critical tasks, inspection dates, and a completion date. As the project progresses, the schedule must be updated to reflect either gains or delays in the schedule and the corresponding end date.

By contract, your general contractor is responsible for the schedule. Keep in mind that you, as the owner, may also share joint responsibility if you need to approve products for the project prior to them being ordered. You might also need to provide utility companies with applications or deposits, hire inspection agencies, furnish items that the contractor must install, respond to any unforeseen existing site conditions, make financial decisions, or you may need to do other various "owner tasks." If the architect is working as a separate entity and is not working for the contractor, they must respond to questions in a timely manner, otherwise they could delay the schedule (questions on a construction project are called RFIs—Request for Information). They must also approve samples of products to be installed in the building prior to the contractor purchasing them; these are called Submittals and are discussed below. Who is responsible for the schedule? The entire project team!

With respect to your schedule, long lead items need to be identified. These items are often specified on projects by designers simply as a means to find out that the end date of a small project may be eight weeks away from the start date, yet the product actually takes six weeks to arrive and would need to be installed during the fourth week of the job to maintain the schedule. A great general contractor who is working on a project for which he is hired by an owner who already has an architect will review the lead time (Lead time: time duration from the time the product is ordered until it arrives at the site, ready for installation) and point it out to the owner and designer. A great design-builder will choose items that fit the schedule, initially avoiding the EWAs/Change Orders.

BASIC BUILDING BIT: *Demand a Critical Path Method schedule with logic ties, and understand the role that you, as the owner, play in schedule delivery. Read and understand the schedule, and have a weekly or bi-weekly schedule update review with your general contractor. Go the to the project and, on a given date and time, look at the date on the schedule and compare what is physically happening on-site to what is stated. Are we on schedule?*

CHANGES

CHANGE ORDERS, CONTRACTOR PHILOSOPHY

"Change orders"—a four letter word times three! (yes, there are twelve letters; did you count?). You need to understand what the different types of change orders are on a project and what your contractor's philosophy is with regard to them. If you're dealing with the employee of a large scale general contractor, it's really important that you understand their philosophy from the top down. During your contractor selection interview, ask them, "What is your philosophy on change orders? How is your project team bonused?" If the only source of employee bonus is enhancing the bottom line on their individual project, you may be setting yourself up for an expensive experience. Beware the low bid contractor that will double their fees at your expense during the course of your project. I do the same research with subcontractor selection.

When are change orders expected and acceptable? Change orders should only be submitted to an owner when an owner requests a change that adds scope, or when there is a true unforeseen existing condition that of which the GC had no prior knowledge. It is also applicable if there is a deficiency in the plans on a project that the owner chose to "hard bid" and not negotiate, or when an official from a building department or other authority having jurisdiction is mandating a change that adds scope and costs to a project. Other than those items, there is no reason to ever accept a change order.

SCHEDULE IMPACT

Any change order should also address the time it takes to add the scope of work to the project and its impact on the CPM schedule. If that cannot be quantified at the time of the change, it should noted that it is being quantified and will be addressed in a future change order. The change order form should state the number of days to add or deduct due to the change.

THERE ARE FOUR TYPES OF CONSTRUCTION CHANGE ORDERS:

Owner Directed—An owner directed change order is an item that the owner desires to add to the project, and he expects that it will cost something.

Existing Condition—An unforeseen existing site condition such as a hidden underground utility line or items behind walls, ceilings, or buried in slabs that are discovered during the building process will in fact cost additional money to remedy. These are legitimate added costs to the project that an owner needs to be prepared to pay for out of their contingency budget.

Inspector Imposed—An "Inspector Imposed" change order is an item on the project that is added to the scope of work that the inspector is requiring to be added to the project. Such change orders may be due to their interpretation of the code, or a local amendment to the code, that must be installed in order for the inspector to approve the project and ultimately allow a final building inspection to pass.

Subcontractor Driven—A subcontractor driven change order is a change order sent to the general contractor for an item that may be deficient in the plans. Some are legitimate, others are not. It is up to a great GC to manage these requests on behalf of the owner.

WHAT SHOULD THE COR DOCUMENT INCLUDE?

A change order document (COR) should include the direct estimated cost to perform the change, any general conditions (indirect work such as dump fees and supervision) necessary to carry out the change, insurance costs, and contractor's pre-negotiated overhead and profit (OH & P). Typically, at the time of this writing, OH & P is typically 10% of the cost of a change order. Depending on the complexity of the project, you may wish to review backup documentation, including subcontractor costs, man hours, labor, equipment, and other documents to support a change.

Occasionally—and this is typical in complex situations, or where there are a lot of unknown items—your contractor may opt to provide the change based on the actual time and materials that it took to complete the change. You should typically authorize a

Time and Material change order in advance of the work being done and receive a "not to exceed" budget for a time and material change.

IMPORTANCE OF A TIMELY REVIEW

In some states, a statute exists that says that an owner has a certain period of days to review and respond to a change order in writing; otherwise, it is accepted. It is important to understand this for each state in which you are operating. A timely response to a change order is very important; a non-response can be legally interpreted as an approval!

SUBMITTAL PROCESS

WHAT IS A SUBMITTAL? A submittal on a construction project is a document and a sample that simply answers the question, "Is this what you really wanted?" Now there is a form for this and a review process that will be dictated in the prime contract and in subcontracts for submittals (OK, definition time again: A prime contract, or general contract, is the legal document that binds the owner and the general contractor. In addition, a subcontract binds the specialty contractors working for the general contractor to the general contractor and to the prime or general contract!).

In the submittal, typically the subcontractor for the specific scope of work reviews the plans and specification and understands what is requested on the project. They then send material samples and product data sheets or drawings to the general contractor. The GC reviews the data against the plans and specifications to ensure that it is correct. The GC then submits this information to the design review professional and/or owner. Upon receiving the signature of the designer and/or owner, the submittal is either rejected, approved, or approved as noted. At such time the submittal is approved, the products are ordered for the project, and upon arrival at a scheduled time, they are then installed.

I recommend submittal review meetings on fast track or remodel projects of short duration. I like to acquire as many submittals as early on in a project as possible. I review them all, then set up a meeting with the design team and the owner and go through each item, each product sample, and acquire approvals during a meeting. Sometimes, this can be a four or eight hour meeting, but we walk out the door with several submittals approved and a project that has congruence and is on a fast track. A lot of time and money can be wasted with submittals being mailed back and forth or not understood. Going through them with the team achieves quick timelines and allows you to manage expectations. Decisions are made in a timely manner and money is saved.

BASIC BUILDING BIT: *A submittal simply says, "This is what you said you wanted on the plans—here is a sample and technical data. Are you sure this is what you really want; and if so, can we order it and install it? Ensure that your schedule accounts for the timeline for submittals."*

Requests for Information

THIS IS A FANCY PHRASE for a question on a construction project! Formal questions on a construction project are put on a form and referred to as a Request for Information (RFI). An RFI should intelligently ask a question, and if possible, reference the plans as well. In addition, the owner/architect should be made aware of any potential cost impacts, for which the general contractor should provide a suggested solution. These questions need to be numbered and kept on a log so they can be kept track of. Any time delays associated with the question should be noted. Additionally, the owner/architect should respond to RFIs with a sense of urgency in order to get the completed answer to the GC so it does not impact the schedule. Have I said this before?—Time is money! And you, the owner, need to move in! RFIs are often transmitted via email, and in the digital age, if there is an existing site condition, sending a photo along with the RFI can greatly increase the speed and ease of the design team understanding the issue and responding with an accurate answer. Imagine trying to do this on a cross-country or transcontinental basis thirty years ago. Some of us have! Thank God for digital cameras and email.

BASIC BUILDING BIT: An RFI says, "We have a problem, or a conflict, and it might cost money; we may have a suggested solution, but we need input from the engineer or owner, and we need it quick. . . ."

Building Inspections

Relationships With Inspectors

It's really important that relationships with building inspectors are honorable. This means the GC has a reputation of doing excellent, code compliant work without taking shortcuts. Inspectors look closely at the plans and specifications, building code, existing conditions, and quality installations. Great GCs do not call for inspections until they know they are ready to pass inspections, and they respect the time of the inspectors. On rush projects in certain jurisdictions, overtime inspections can be requested for a fee. It's imperative that the GC ensure that the approved plans (i.e.—plans approved by the building department) and the work installed correspond. They should be identical.

Building Basic Tip: Always treat building officials with the utmost respect. It makes the project go well, and inspections get signed off quicker!

QAA Inspections

What is QAA?—A Quality Assurance Agency is often required in certain jurisdictions. This is really a term for a special inspections and testing company. Many items on a project will require this inspection service, in addition to the inspections performed by the building department. The QAA is a firm that will measure and test items such as concrete, welds in place, bolting, rebar, and compaction to ensure that what is installed matches the engineered plans. In the case that there is a deficiency between the plans and specifications, a noncompliance report (NCR) will be issued by the QAA. The general contractor must then take appropriate steps or work with the architect/engineer of record to ensure that the NCR is "cleared." It's a very important part of each project that is not to be missed. Additionally, this QAA cost must be incorporated into your project pro forma if it is required in the area of your project. Ask your contractor, architect, or design-builder if QAA/Special Inspection is required on your project, or you can also verify it with the local building department.

BASIC BUILDING BIT: Check with your GC, or, better yet, have a category for NCRs in the Owner-Architect-Contractor meetings, and ensure that the NCRs are being "cleared" in a timely fashion, otherwise, it can hold up your Certificate of Occupancy.

PROGRESS PAYMENTS

TIMELY PROGRESS PAYMENTS are a huge advantage to an owner. Why? I call it the "payment trickle down effect." The payment trickle down effect on a construction project is the impact that timely or untimely payments have on the bottom line of the project. When an owner pays in a timely manner, the general contractor is also able to release payments to subcontractors and suppliers in a timely manner. This results in goodwill from all the subcontractors and suppliers indirectly to the owner.

How so? On projects where funds are dispersed on time, subcontractors are often able to pay their suppliers more quickly, and they are able to take advantage of cash discounts. When subcontractors are paid on a timely basis, they realize, and often express, "this is a good owner," and they will often mitigate or not charge for small odds and ends that come up. Time and again, I have seen that when subcontractors are paid quickly, they respond more quickly to requests. In other words, as an owner, if you can make it a priority to honor your payment schedule or do better than it, you will build goodwill with the entire project team and this results in time and money savings to your bottom line.

When you review your general contract, the payment terms and conditions should be spelled out in detail. For most commercial building projects, expect to make progress payments for work as it is installed on a monthly basis. Progress inspections can be performed by you, your banker (if a loan is involved), or an architect to ensure that the percentage of the work being billed accurately reflects the installation. Special deposits or payments for off-site materials may be required in order to maintain an aggressive construction schedule, so also ensure that you understand this expectation. If an owner makes payments for off-site materials, your general contractor should provide documentation that indicates that the items were purchased for the project, are stored in an insured and bonded facility, and that title has transferred to your project.

Another increasingly popular and efficient method of project payment administration is the use of a third party disbursement account.

What is a construction disbursement account? This is a third party company that is hired by the owner to provide an account that serves as evidence of funding for the project and performs disbursement of progress payments for a reasonably low fee. Some states, at the time of this writing, have passed laws requiring that this type of firm is hired on certain construction projects. Be sure to check with your contractor or the state in which you are building to find out if this is a law.

Your disbursement agency will inspect progress of the work installed to ensure that you are paying out progress payments on time and accurately. They will gather and review various lien releases, labor releases, and other legal documents to ensure that the project is free of encumbrances. They will ensure that payments are not released to vendors without releases, and they typically set your funds up in an interest bearing account.

We often encourage clients to utilize these services in today's market where possible. Why? Well, not only due to the benefits we already discussed, but again, better pricing as well. How is this so? When a general contractor announces a project bid and can state that the project is fully funded through a disbursement account, the subcontractors and suppliers are then confident that money and payments will be a non-issue. The benefit to the owner is that a third party is now "watching" the progress of the project, ensuring funds are only paid for work that is provided, the project is free of encumbrances, and subcontractors and vendors are paid in a timely manner. This results in a smooth project administration, and also results in bottom line savings to an owner.

More Basic Building Bits

As-Builts

An As-Built is a final set of plans that shows the conditions of the project after construction. It is simply a marked up set of drawings that show any minor revisions to the plans that were made or required as the project was built. It is literally as it was built—thus, the term "As-Built." You will want to get a full hard copy of these plans, in addition to an electronic file of the plans.

If your architect has made revisions, at the end of the project, you can also ask the architect for the electronic files in CAD and PDF; that way if you need to do any future design or construction or sell your building, you will have all the documents you need already. It is very difficult for future projects and for maintenance companies to perform work in future years without having this information, and often, unfortunately, sometimes the original building contractor or architect may not be in business years after your project is completed. We recommend getting paper documents, electronic PDFs and CADs, or the latest design files.

Authorities Having Jurisdiction

How many agencies can be involved in a single construction project? A lot! Navigating your way through these requirements leads to financial success or failure, and again, timeline issues. Your design-build general contractor, or general contractor working in conjunction with your hired design professional, must know the requirements of each authority that has jurisdiction over your project; and, more importantly, have relationships with the people that work for these agencies in order to expedite your work. On a single construction project, you will likely deal with Planning, Zoning, Sanitation, Building Departments, Fire Departments, the Health Department, State Fire Marshal, Air Quality Management, and OSHA, just to name a few.

When interviewing your team prior to starting the project, understand their strength and knowledge of the processes, and timelines in dealing with these various agencies. As an owner, other than understanding the costs of fees and permits with these agencies that you will need to account for and build into your pro forma, you should rarely have to interact directly with these agencies if your design-builder is a great contractor.

BUILDERS RISK

Often, a construction project is adjacent to or within another building, or near property that can be damaged. For example, if you have a tenant suite under construction in a building that has multiple suites, and your project catches on fire as a result of the construction in your suite, you can be held liable for damage to the shell building, the adjacent tenant spaces, and other items of value. It is important that a Builders Risk Policy is addressed in your Prime Contract and that either you, as the owner, have procured such a policy, or that the general contractor has this policy in place. Always ask for a copy of this policy, read the fine print and exclusions, and make certain that the deductible makes sense. A large deductible may not fully protect you. Typically, deductibles and rates are inversely proportional.

BUILDING VALUE/SUSTAINABILITY

I had the pleasure of serving on the development team of the first LEED Gold office building in Las Vegas. It was a multi-story tower. At the time, we were frontrunners. We attended classes and hired architects that were educated in the LEED and green building process.

Once you have done it and understand the process, it really isn't that difficult to obtain a LEED certification—it's based on a point system applied to the design and construction process. Design per the LEED requirements, administrate the construction per the documents and LEED system, and it results in a LEED building. The future of the built environment is dependent on the use of sustainable practices during the building process, which requires you to use recycled, sustainable, and environmentally friendly building products in the building; and energy should also be used wisely. Contact your local chapter of the United States Green Building Council (USGBC)_at USGBC.ORG. I can tell you that the future of building code is going to incorporate a lot of the LEED principles into building code so that everyone will be building green and sustainable in future years to come. It makes sense to me.

GENERAL CONTRACTOR RECORD

A Clean Site—Keep it Clean! A clean site makes for a safer environment. Your GC must have a clean and organized working environment. Subcontractors should, by contract, be held to a high standard of daily site cleanup in order to maintain a safe, hazard free environment. Just reducing tripping hazards alone can prevent loss of time and injury to employees.

INSURANCE FOR CONTRACTORS

It wouldn't be very fun to wake up and find out that there is a claim on the project, and that the contractor does not carry proper insurance. It is very important that general liability insurance is in place on all operations assigned to the GC and that the insurance certificate provided specifically names the owner as additionally insured.

LIABILITY

Liability for safety is a big issue that can cost every company involved substantially. In California case law, if an employer willingly and knowingly causes an employee to perform an unsafe activity and the employee is injured or killed, the employer is subject to not only civil litigation, but can also be criminally prosecuted. Additionally, authorities having jurisdiction, such as OSHA, can levy heavy penalties and fines for safety violations.

O & M MANUALS

An O & M Manual is a guide book that shows different systems, information on equipment, cleaning information, and maintenance information that will all be neatly and orderly placed into binders and provided by the general contractor at the completion of the project.

ON-SITE SAFETY: IMPORTANCE TO OWNER? PEACE OF MIND

Construction activities involve people performing dangerous activities on a daily basis. It's very important that the contractor provides safety measures at your site. We always remember that while the project schedule is important to maintain, providing a safe site, where workers can go home safely to their families, is of utmost importance. Safety is the first priority above everything else on our projects. Your chosen GC should have a site-specific safety plan, require weekly on-site safety meetings, and require all subcontractors and vendors to provide safety documentation. As an owner, when you arrive at your jobsite, you should see proper safety notices, a first aid center, an MSDS center (this is a list of all hazardous items on the project, and what actions to take if exposure occurs), signage showing the location of the nearest hospital, and notices to dial 911 in case of an emergency.

ON-SITE SECURITY

When hiring on-site security, it is important to review of their contract, as well as understand their insurance policy. Make sure they have some responsibility in an event resulting in loss.

PUNCH LIST

A punch list is typically done at or near project completion, or, on very large projects, it is done by area as each phase or section of a project is completed. Typically, the owner, architect, and general contractor walk through each room of a building and look for any items that appear to be flawed or incomplete. A detailed list of each item, listed by location and trade, is then written by the general contractor and distributed to the subcontractors.

As an owner, it is important to understand that you should allow the contractor enough time to remedy any punch list items. Once the items are remedied, the owner, architect, and contractor will walk the project again and check off the fact that each item is now complete and acceptable. Some owners move in to their space prior to the completion of the punch list; this is an error, because it becomes difficult to complete the punch list items in an occupied space. The lines of responsibility can become blurred if you, as the owner, made any damage to the space during your move-in. Typically, a general contractor will have a statement in the contract that states that there shall be one punch list; and if the owner moves in prior to the completion or during the compilation of the punch list, the owner accepts the space "as-is." A great general contractor will typically perform quality checks and pre-punch list walks throughout the construction process; and the owner, architect, contractor final punch list should only have minor items to address. As an owner, keep quality in mind throughout the project; and ensure that you schedule time near the end of the project to be available to participate in the creation of the final punch list and sign off on it prior to moving in to your project.

THEFT AND VANDALISM

A Builders Risk policy can also protect against theft

and vandalism at the site. Unfortunately, in today's world, theft and vandalism are very common on construction projects, and measures against theft must be taken, such as employing on-site security.

TRAINED SUPERVISION

Ensure that your GC has supervision on-site that complies with federal and state standards and possesses OSHA certifications for on-site safety, basic first aid, etc. It is required by law in many states, but most owners do not think to ask this question.

TRAINING

In addition to the As-Builts, O&M manuals, and warranty information, certain systems may require basic training. Audio-Visual systems, HVAC systems, or medical gas systems may require the installation company to provide you with a few hours of training. Often, this training can be video recorded for record and future reference. Please check with your design-builder, architect, or general contractor at the beginning of your project to determine or request specific training.

UNIONS

Since construction involves labor, there are several unionized crafts. Depending on your geographic location, project size, and project funding, union labor may or may not be required on your project. Political positions on unions are sometimes controversial; but trade unions do provide a source of trained, educated workers that specialize in a particular craft. Typically, the average union tradesperson has experienced several years of technical trade school and several thousand hours of supervised on-the-job progress training prior to being allowed to receive the title of "journeyman." If you plan a project in a heavily unionized area, the benefits of using union labor may outweigh the headaches or risks involved in not doing so. Again, your general contractor should be a source of advice with regard to this decision.

WARRANTY

Typically, the warranty on a building is one year. Each subcontractor warrants their scope of work for material and labor for one year after substantial completion. In your close out documents, you will have a written warranty from each subcontractor who performed work on your project.

There are a few key warranties to be aware of that are most important. Roofing, if applicable, typically has a longer period than one year. It is sometimes specified by the architect or can be decided upon by the owner, but you should consider getting a good warranty on your roof. Warranty periods on the roof can be purchased for up to twenty years. A decision about how long you will hold the building as an asset and the future sale value of the building need to be considered when deciding on how long a roof warranty you would like to buy. Be sure to discuss this item with your architect and general contractor. A leak free roof creates happy occupants!

If your project has landscaping, ensure that the one year warranty on the plants covers replacement of the plants should they perish. It can be very costly to replace dying plants. Often having a good landscape maintenance agreement in place by a professionally licensed landscaper, preferably the firm that did your landscape construction and installation, will save a lot of headaches. Some cities require that this agreement be in place before they will grant you a certificate of occupancy.

BASIC BUILDING BIT: *Get a good warranty on your roof. It will add value to your building and ensure that your roof is leak free!*

Pulling it All Together

Wow! Are you tired yet? Hopefully, we haven't bored you to sleep. If you are still with us, let's pull it all together.

Pro Forma

The biggest challenge that owners often face is pulling it all together. Often, I find that owners are not aware of many costs involved in a project that are NOT included in the general contractor's price, because either it is typically not included, or, for whatever reason, the general contractor simply cannot include it. A great general contractor will send a qualified, detailed narrative as the proposal that should point out "Other Costs" that you need to consider, as well as a list of "Exclusions." Both items need to be paid close attention to so you can manage your overall budget. The general contractor's price for construction should be one of several numbers that you are keeping on a spreadsheet, which is the budget/pro forma for the overall project.

Consider soft costs, which are any costs that are not in the general contractor's construction contract, and be sure to include these costs in your budget—including design fees, consulting fees, taxes, sales and marketing expenses; fees to authorities having jurisdiction, such as planning, zoning, water districts, sanitation districts, and meter fees; finance charges, interest carry (debt service) on a construction loan, and travel to the project, if necessary. You should carry reasonable, accurate estimates based upon real input by identifying all the various potential expense items; and then also add a soft cost contingency of 5-10% (recommended) of your total.

In addition to the soft costs, you will have your various hard costs. Budget for this include the cost to install utilities, signage, anything excluded from your general contractor's quote, furniture, fixtures, equipment, and items needed at move-in.

BASIC BUILDING BIT: Your contractor's price is only one cost in your total project. Ensure that you do a detailed pro forma and capture all the costs!

CONSTRUCTION – DAILY PROGRESS

The daily progress on a construction project should follow a Critical Path Method schedule. Often, there can be delays due to unforeseen existing site conditions—especially if you have utility or underground work in your project. When issues that can potentially delay the progress of the project are discovered, your general contractor should immediately notify you, the owner, and the architect of any delays. One thing that an owner must keep in mind, is that the General Conditions costs for the general contractor (including the cost of a supervisor and staffing) is a per day cost that is incurred even when the site is not productive. The owner should expect to pay for the General Conditions for any schedule extensions, even if it is due to an unforeseen site condition outside the control of all parties.

I heard a conversation between a client and his GC regarding a delay. The client had experienced delays to the project due to buried underground electrical lines. It took two weeks to achieve a resolution from the power company on these lines. Unfortunately, the building pad could not be completed until the issue was resolved, and the building pad is, obviously, a critical path activity. It must be completed within a specific timeframe, or, for each day that it takes longer to complete, the schedule will be extended by at least that day or even exponentially. The owner said, "But how can there be a delay? I saw men working on the other side of the site?" The logic here must be clearly understood; if the electrical issue was holding up the completion of the building pad, it wouldn't matter if there were fifty workers sweeping the parking lot or digging a trench that was not a critical path item.

As the owner, understand and read your construction schedule and clearly understand where the critical path is. Understand that while delays are not welcomed by anyone, and it is a huge annoyance to a productive general contractor—often meaning his project management team is scrambling for the fastest resolution—the cost of delays and time extensions is a legitimate cost that you, as the owner, will be expected to pay. Therefore, on the pro forma we touched on in a prior section, always carry a hard cost contingency of your total hard costs so you will have funds in place to pay for unforeseen conditions or delays. It is just reality in construction. If you expect it, then it won't be a financial surprise when it happens; and if it doesn't happen, then you will have a budget surplus which is much better than a deficit! Proper financial planning is the key to your project's success!

BASIC BUILDING BIT: *You will have budget overruns in some area of your total project. If you don't, be pleasantly surprised. Always carry a contingency!*

OAC MEETINGS

Depending on the project size and scope, it may be necessary to have a weekly Owner-Contractor Meeting or Owner-Architect-Contractor (OAC) meeting. Most meetings should take about an hour to two hours. Meeting notes with a ball in court/responsibility section and deadline should be maintained for each open issue. The meetings should cover: permits/drawing revisions; issues with any inspections; review of any requests for information; review of submittals and long lead items; review of change orders; review of project schedule; review of or tracking of progress payment timeline; safety; quality; punch list; and any other business. The meeting minutes from each meeting should be utilized as a tool by all the parties at the meeting to ensure that each member is completing their responsibilities prior to the next meeting, and that the project is being pushed to a completion.

CLEAN UP

Upon completion of the project, the general contractor will employ a final cleaning firm to clean your space top to bottom. It is important to ensure that cleaning requirements are clearly defined at the beginning of the project. Also, after the owner moves in, the owner should not expect the contractor to continue to provide additional cleaning trips. The final cleaning should be completed with the final punch list and the space turned over to your complete satisfaction before moving in. Wow! We are almost done!

MOVE-IN

During the move-in process, a prudent owner will go through the space and ensure that everything is working well. If you have a warranty issue, it's best to discover it as soon as possible so that subcontractors will quickly mobilize and take care of any warranty items. Use everything within the first week, and ensure that you are happy.

CONCLUSION

Did you have fun? Are you having fun? While there are a lot of details, parts, pieces, plans, companies, and people involved in even the smallest construction project—if you keep *Your Little Black Book of Building Basics* close by, we hope that it will help you to make your construction project a success, and the overall process one that you enjoy!

Resource List

ABC—Associated Builders and Contractors
4250 N. Fairfax Drive, 9th Floor
Arlington, VA 22203-1607

Phone: (703) 812-2000

Website: www.abc.org

AGC—Associated General Contractors
2300 Wilson Boulevard, Suite 400
Arlington, VA 22201

Phone:
(703) 548-3118 - Local
(800) 242-1767 - Publications

Website: www.agc.org

AIA—American Institute of Architects
1735 New York Avenue NW
Washington, DC 20006-5292

Phone: 800-AIA-3837 or 202-626-7300

Website: www.aia.org

CCIM Institute
430 N. Michigan Avenue, Suite 800
Chicago, IL 60611

Phone: (312) 321-4460

Website: www.ccim.com

ICSC—International Council of Shopping Centers
1221 Avenue of the Americas, 41st Floor
New York, NY 10020-1099

Phone: (646) 728-3800

Website: www.icsc.org

NAIOP, The Commercial Real Estate Development Association
2201 Cooperative Way, Suite 300
Herndon, VA 20171-3034

Phone: (703) 904-7100

Website: www.naiop.org

SIOR—Society of Industrial and Office REALTORS®
1201 New York Avenue NW, Suite 350
Washington, DC 20005-6126 USA

Phone: (202) 449-8200

Website: www.sior.com

ULI—Urban Land Institute
1025 Thomas Jefferson St. NW, Suite 500 W
Washington, DC 20007

Phone: (202) 624-7000

Website: www.uli.org

USGBC—U.S. Green Building Council
2101 L Street NW, Suite 500
Washington, DC 20037

Website: www.usgbc.org

About the Author

Jeff P. Manning is a leading author, multi-licensed general contractor's qualified manager, and real estate development consultant. Jeff also makes himself available as a consultant to other general contractors, and assists developers in general contractor selection and project management.

With more than two decades of industry expertise, Jeff founded a highly successful contracting organization in the western United States. Jeff has personally overseen projects on Las Vegas Boulevard and throughout the western U.S. for ground up construction and interior remodels. He has helped build millions of square feet of buildings in various product types. Jeff has also helped develop several NAIOP Spotlight award-winning projects.

In addition to his work as a general contractor, Jeff has authored several books, has been a guest columnist in local publications, and supports industry organizations such as the Associated General Contractors, NAIOP The Commercial Real Estate Development Association (including serving two years on the NAIOP board), American Institute of Architects, and other organizations in real estate development, construction, and architecture through time and financial sponsorships.

Jeff was around real estate and construction as a child, was valedictorian of his high school graduating class, and earned his bachelor's degree in business administration and construction management. In his early career he was mentored by some of the most talented, demanding industry leaders.

If you need assistance with project task check lists, pro formas, consulting, design, or general contracting, or you need to build a project and don't know who to turn to for advice, Jeff Manning may be reached via email at jeffpmanning@jeffpmanning.com, or visit the last section of the book for updated contact information.

CPSIA information can be obtained at www.ICGtesting.com
Printed in the USA
BVOW01s1920210916

462917BV00006B/11/P

9 780615 352893